Guy de Maxence AFANDA

MATHEMATICAL
DELIGHTS

THE THEOREM OF DE FERMAT

Let us remember this theorem:

for n>2, $\nexists$ (x, y, z) $\in N^*$ / $z^n = x^n + y^n$, otherwise, for n upper than 2, there does not exist three natural number different of zero so that: $z^n = x^n + y^n$.

Using calculation of probability, we proceed as:

1) For n=0, the probability to verify the theorem does not exist
2) For n=1, the probability to verify the theorem is total as calculated so:

$$p = \frac{z - x + z - y}{z} \quad =1$$

3) For n=2, the probability to verify the theorem is:

$$p = \frac{(z - x)^2 + (z - y)^2}{z^2}$$

Using the quadratic trio (2, 3, 5) as follows: $z^2 = x^2 + y^2$
$\equiv 5^2 k^2 = 3^2 k^2 + 4^2 k^2$, we get:

$$p = \frac{(5k - 3k)^2 + (5k - 4k)^2}{25k^2} = \frac{5}{25} = \frac{1}{5} \; ; k \in N^*$$

We can prove it by the following table:

3

1	4	9	16	25(k=1)
49	36	64	81	100(k=2)
144	121	169	196	225(k=3)
289	256	324	361	400(k=4)
484	441	529	576	625(k=5)

4) For n>2, the probability to verify the theorem is:

$$\begin{cases} p = \dfrac{(z-x)^n + (z-y)^n}{z^n} \\ z = \sqrt[n]{x^n + y^n} \end{cases}$$

For the general accuracy in checking the theorem let us notice that:

$$0 \le pz^n \le z^n$$

So:

$$0 \le (z-x)^n + (z-y)^n \le z^n.$$

For n=2, we reach to the result: $z \le x + y$.

CALCULUS

ANALYSIS

Change is measured by:

$\Delta x = x_h - x_0 = x - x_0$

By extension:

$\Delta^2 x = x_{2h} - 2x_h + x_0 = x_2 - 2x_1 + x_0$

In general:

$$\Delta^n x = \sum_{i=0}^{n} C_i^n (-1)^{n-i} x_{ih}$$

Therefore, if f is a function defined as f(x), then:

$\Delta f(x) = f(x+\Delta x) - f(x)$

$\Delta^2 f(x) = f(x+2\Delta x) - 2f(x+\Delta x) + f(x)$ $\Delta^n f(x) =$

$$\sum_{i=0}^{n} C_i^n (-1)^i f[x + (n-i)\Delta x]$$

By extension:

$\Delta f(x,y) = f(x+\Delta x, y+\Delta y) - f(x,y)$

$$\Delta^2 f(x,y) = f(x+2\Delta x, y+2\Delta y) - 2f(x+\Delta x, y+\Delta y) + f(x,y)$$

$$\Delta^n f(x_1, x_2, \ldots, x_m) =$$

$$\sum_{i=0}^{n} C_i^n (-1)^i f[x_1 + (n-i)\Delta x_1, \ldots, x_m + (n-i)\Delta x_m]$$

Differentials are:

$df = f'dx$

$d^2f = f''dx^2$

$d^nf = f^{(n)}dx^n$

By extension:

$df(x,y) = f'(x,y)(dx,dy)$

$$\begin{cases} f(x,y,z,\ldots) = f(x_1, x_2, \ldots, x_n) \\ df = f^{(n)}(dx_1, dx_2, \ldots, dx_m)^n \end{cases}$$

Then let us talk about divariation, the change of a function according to two simultaneous changes in to two different systems of change analysis. We notice it as:

$$\tilde{\Delta} f(x) = f(x) - g(x)$$

Then:

$$\Delta\tilde{\Delta}f(x) = \tilde{\Delta}^2 f(x) = \tilde{\Delta}f'(x)\Delta x$$

It is useful when the function or the quantity is studied according to two different standard of reference simultaneously.

Digression is:

$$\tilde{\nabla}f(x) = \frac{\Delta f(x)}{\Delta x} - \frac{df(x)}{dx}.$$

INTEGRATION

Let us begin we: $f''df=f'df'$, equivalent to: $df=\lambda df'$, where λ is the pace of recurrence. When λ is simple, that is if : $d^n\lambda=0$, is available, then there exists an analogical relationship accurate that gives the link between f and f', and then between F the primitive of f and f.

For example:

i)   When: $d\lambda=0$; calculation leads to: $f''^2=f'f'''$; and by analogy: $f'^2=ff''$, $f^2=Ff'$

ii)   When: $d^2\lambda=0$; calculation leads to: $2f'f'''^2-f'f''f^{IV}-f'''f''^2=0$; by analogy: $f(2f''-\lambda f''')=f'^2$, $F(2f'-\lambda f'')=f^2$

when λ is not simple, calculation gives to remember that yet: $f''\lambda'=f''-\lambda f'''$. Two ways lead to determine λ'' as:

1) $f''^2\lambda''=-\lambda'f''f'''-\lambda f''f^{IV}+\lambda f''''^2$, which by analogy leads to:
$$\begin{cases} f^2\lambda'' + ff'(1+\lambda') = \lambda f'^2 \\ F^2\lambda'' + Ff(1+\lambda') = \lambda f^2 \end{cases}$$, containing the cases: $\lambda'=0$, and: $\lambda''=0$

9

2) $f''\lambda''=f'''(1-2\lambda')-\lambda f^{IV}$, which is useless for analogy.

ALGEBRICAL EQUATION WITH ONE UNKNOWN
ELEMENT

Let us write them in the general mode, that is:

$$\sum_{i=0}^{n} a_i x^{n-i} = 0$$

, where x is the unknown element and a_i are the data.

For the resolution, practice shows the relevance of the following writing:

$$a_0^{n-1} \sum_{i=0}^{n} a_i x^{n-i} = (Ax + B)^n - C = 0$$

Where A B and C are determined according to the equation

Let us for examples, take both cases: n=2, n=3

1) n=2; we have: $a_0 x^2 + a_1 x + a_2 = 0$; then:

$a_0^2 x^2 + a_0 a_1 x + a_0 a_2 = |Ax+B|^2 - C = 0$

Calculation leads to: $\begin{cases} A^2 = a_0^2 \\ 2AB = a_0 a_1 \\ B^2 - C = a_0 a_2 \end{cases}$, then:

$\begin{cases} A^2 = a_0^2 \\ 4A^2B^2 = a_0^2 a_1^2 \\ C = B^2 - a_0 a_2 \end{cases} = \begin{cases} A^2 = a_0^2 \\ 4B^2 = a_1^2 \\ 4C = a_1^2 - 4a_0 a_2 \end{cases}$, and the

equation become simply:

$$|2a_0 x + a_1|^2 = a_1^2 - 4a_0 a_2$$

2) n=3; then: $a_0 x^3 + a_1 x^2 + a_2 x + a_3 = 0$, and:

$a_0^3 x^3 + a_1 a_0^2 x^2 + a_2 a_0^2 x + a_0^2 a_3 = (Ax+B)^3 - C = 0$; whence:

$\begin{cases} A^3 = a_0^3 \\ 3A^2 B = a_1 a_0^2 \\ 3AB^2 = a_2 a_0^2 \\ B^3 - C = a_3 a_0^2 \end{cases} =$

$\begin{cases} A^3 a_1^3 = a_0^3 a_1^3 \\ 3a_1^2 A^2 B = a_0^2 a_1^3 \\ 3a_1 AB^2 = a_2 a_0^2 a_1 \\ B^3 - C = a_3 a_0^2 \end{cases} = \begin{cases} A = a_0 \\ 3B = a_1 \\ (Aa_1 + B)^3 = a_0^3 a_1^3 + a_0^2 a_1^3 + a_0^2 a_1 a_2 + a \\ \qquad\qquad + C \end{cases}$

In the general way, a mean of resolution is:

$$\begin{cases} A^n = a_0{}^n \\ nB = a_1 \\ (Aa_1 + B)^n = a_0{}^n a_1{}^n + a_0{}^{n-1} \displaystyle\sum_{i=0}^{n-1} a_1{}^{n-1-i} a_{1+i} + C \end{cases}$$

The equation takes therefore the form:

$(na_0x+a_1)^n = n^nC$; then:

$$(na_0x+a_1)^n = a_1{}^n(na_0+1)^n - n^n a_0{}^n a_1{}^n - n^n a_0{}^{n-1} \sum_{i=1}^{n} a_1{}^{n-i} a_i$$
.

THE IMPROVED PROBABILITY

Let us here formulate the rate of certainty c as:

$$c = \frac{\text{the countable amount of an expected event } e}{\text{the amount of all possible events}} = \frac{n_e}{card\{e,...\}}$$

Let us also formulate the rate of uncertainty i as: $i = 1-c$. therefore, by the mean of n occasions, that is n times of trying to get the event expected, the probability to get it is: $p = 1-i^n$; it increases when $i \neq 1$.

After, with respect to the probability of the incompatible event: $\bar{P}=1\text{-}p$, let us have:

1) A and B two separate expected events, $p_{A,B}$ the probability to have both A and B, as:

   $$p_{A,B}=n_{A,B}\,p_A p_B$$

2) The probability to have one without the other, as: $p_{A/B}=1\text{-}p_{A,B}$

3) The probability to have A before B, as: $p_{A;B}=$

   $$\frac{p_A}{p_A+p_B}=1\text{-}p_{B;A}$$

4) The probability of A according to the probability of B (or conditional probability), as:

   $$p_{A\supset B}=p_{(A;B),(A,B)}=p_{B;A}\,p_{A,B}\;.$$

THE EXACT VALUE OF π

Let us observe the following figure:

There is the circle and the inner lozenge. The circle is covered along: s=Rα, and the inner lozenge as: e=Rα√2

Further, let us also consider the following figure:

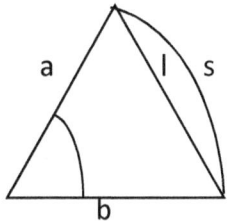

Where:
$$\begin{cases} l = \sqrt{a^2 + b^2 - 2abcos\alpha} \\ e = \sqrt{|a - b|^2 + 8ab\alpha^2/\pi^2} \end{cases}$$

Accurate calculation establishes:

$$s = \frac{e + \sqrt{e^2 + 4R_m{}^2\alpha^2/\pi^2}}{2}$$, as provider of: $\pi = \sqrt{2} + \sqrt{3}$

, as far as R_m is the mean radius (for the circle: R_m=R, and: a=b).

For the measure, the unit of measure for π is so that: $\pi = 3,14...\breve{m}/\overline{m}$ (3,14...meter curved per meter right), as for any Euclidean angle, an angle with two right sides in a plan. Consequences are:

1) The perimeter of the circle is the rotation of the radius along the angle of rotation; then: p=

$$\int_0^{2\pi} Rd\alpha = \pi D$$

2) The surface of the circle is the sweeping of the radius along the half perimeter; then: A=

$$\frac{1}{2}\int_0^p Rdp = \pi R^2$$

3) The surface of the sphere is the run of a half circumference on a circumference; then: A=

$$\int_0^p \frac{1}{2}pdp = \frac{\pi^2 D^2}{4}$$

4) The volume of the sphere is the run of a half area of circle along a half circumference; then:

$$V= \frac{\pi^2 D^3}{12}$$.

STRUCTURAL REPRODUCTIONS

Let us consider that: $C^{(a,b,c)}_2 = (a,b),(a,c),(b,c)$.

Further, $C^{a,b,c}_{x_i + x_j} = a + b, a + c, b + c.$

Hence: $A^{a,b,c}_2 = (a;b),(a;c),(b;c),(b;a),(c;a),(c;b).$

Yet: $\Sigma^{a,b,c}_2 = (a + b) + (a + c) + (b + c)$, and:

$\Pi^{a,b,c}_2 = ab + ac + bc$

These operations can be extended to any structural reproduction.

21

EXPONENTIALS

First, there is the case of a^n, where **a** is a natural number.

$$\begin{cases} 2^n = \displaystyle\sum_{i=0}^{n} C_i^n \\[2mm] 2^n = \displaystyle\sum_{\alpha=0}^{1}\left(\sum_{i=1}^{n} C_i^n \alpha^{n-i}\right) \end{cases} \text{, and therefore:}$$

$$a^n = \sum_{\alpha=0}^{a-1}\left(\sum_{i=1}^{n} C_i^n \alpha^{n-i}\right) = \sum_{\alpha=0}^{a-1}[(\alpha+1)^n - \alpha^n]$$

Further, for a^x where x is any real number, we remember of the integer function E defined as: $E(x)=E(e \leq x < e+1)=e$, where e is an integer number. And for the extension of the factorials to any real number,

$$\begin{cases} x! = \displaystyle\prod_{i}^{E(x)-\lambda} (x - |\lambda|i) \\[2mm] \lambda = \dfrac{E(x)}{|E(x)|} \end{cases}$$

we reach to:

Then precisely:

$$a^x = \sum_{\alpha=0}^{a-1} \left(\sum_{1}^{E(x)} C_i^x \alpha^{E(x)-1} \right)$$

For complementary purpose, every natural number different of zero is a duomerical complex. In other words, each of these numbers is formed as follows:

$$N = \sum_{i=0}^{n} \alpha_i 2^{n-i} \prod_{=j=1}^{m} \left(\sum_{i=0}^{q} \beta_i 2^{q-i} \right)_j$$

$(\alpha_i, \beta_i) \in \{0,1\} \times \{0,1\}$; n and q are duomerical wideness and m is the coefficient of complexity of N. N is a premier number if: m=1. Premier numbers are therefore continuous or saturated (solid) numbers, and the others are discontinuous or unsaturated (fluid) numbers. So:

1) 2 is the (natural) stump of the complexity
2) Every premier number is the root of the function of complexity C defined by:

$C(N)=\text{card } (N=\prod N_i) - 1$

3) Each premier number is then a stake of the application of position P defined by: $P'(N)=C(N)$

Now: $\Delta\text{card } (N=\prod N_i) = \text{card}(N+x=\prod N_i) - \text{card } (N=\prod N_i)$. Here, x=N-p , p is the premier number necessary to N, and then: Δcard $(N=\prod N_i) =$ card $(2N-p=\prod N_i) -$ card $(N=\prod N_i)$.

Hence, P''(N)= C'(N)= $\dfrac{C(N)}{N-p}$. That is: all the premier numbers are independent of each other, in the sense that they are all equivalent according to any statement to which they are reported (any remark made about a premier number as a premier number is valid for any other premier number).

EXTASES
MATHEMATIQUES

LE THEOREME DE DE FERMAT

Le théorème : Pour n>2, $\nexists$ (x, y, z) $\in N^{*}$ / $z^n = x^n + y^n$. Donc pour n supérieur à 2, il n'existe pas de triplet naturel non nul tel que la relation précédente soit vérifiée. Avec le calcul de probabilité nous procédons ainsi :

1) Pour n=0, la probabilité que le théorème soit vérifié n'existe pas

2) Pour n=1, la probabilité que le théorème soit vérifié est totale ; en effet, elle est:

$$p = \frac{z - x + z - y}{z} = 1$$

3) Pour n=2, la probabilité que le théorème soit vérifié est :

$$p = \frac{(z - x)^2 + (z - y)^2}{z^2}$$

En utilisant le triplet quadratique (3, 4, 5) tel que :
$z^2 = x^2 + y^2 \equiv 5^2k^2 = 3^2x^2 + 4^2y^2$, nous obtenons :

$$p = \frac{(5k - 3k)^2 + (5k - 4k)^2}{25k^2} = \frac{5}{25} = \frac{1}{5} \; ; k \in N^*$$

Pour preuve, dressons le tableau des carrés suivant :

1	4	9	16	25(k=1)
36	49	64	81	100(k=2)
121	144	169	196	225(k=3)
256	289	324	361	400(k=4)
441	484	529	576	625(k=5)

4) Pour n>2, la probabilité pour que le théorème soit vérifié est :

$$\begin{cases} p = \dfrac{(z - x)^n + (z - y)^n}{z^n} \\ z = \sqrt[n]{x^n + y^n} \end{cases}$$

Pour la saisie générale, la condition suivante doit être accomplie :

$$0 \leq pz^n \leq z^n$$

Soit :
$$0 \leq (z - x)^n + (z - y)^n \leq z^n.$$

Pour n=2, nous arrivons au résultat : $z \leq x + y$.

L'ANALYSE

La variation est définie : $\Delta x = x_h - x_0 = x - x_0$

Par extension : $\Delta^2 x = x_{2h} - 2x_h + x_0 = x_2 - 2x_1 + x_0$

En général : $\Delta^n x = \displaystyle\sum_{i=0}^{n} C_i^n (-1)^{n-i} x_{ih}$

Aussi, f étant une fonction définie par f(x), alors :

Δf(x) =f(x+Δx)-f(x)

Δ²f(x) = f(x+2Δx)-2f(x+Δx)+f(x) Δⁿf(x)=

$$\sum_{i=0}^{n} C_i^n(-1)^i f[x + (n-i)\Delta x]$$

Par extension:

Δf(x,y)=f(x+Δx,y+Δy)-f(x,y)

Δ²f(x,y)=f(x+2Δx,y+2Δy)-2f(x+Δx,y+Δy)+f(x,y)

$\Delta^n f(x_1,x_2,...,x_m)=$

$$\sum_{i=0}^{n} C_i^n(-1)^i f[x_1 + (n-i)\Delta x_1,...,x_m + (n-i)\Delta x_m]$$

Les différentielles sont alors :

df=f'dx

d²f=f''dx ²

dⁿf=f⁽ⁿ⁾dx ⁿ

Par extension:

df(x,y)=f'(x,y)(dx,dy)

$$\begin{cases} f(x,y,z,...) = f(x_1,x_2,...,x_n) \\ df = f^{(n)}(dx_1,dx_2,...,dx_m)^n \end{cases}$$

Parlons maintenant de la divariation, la variation d'une fonction selon deux variations concomitantes par rapport à deux repères différents ; elle est définie ainsi :

$$\tilde{\Delta} f(x) = f(x) - g(x)$$

Donc :

$$\Delta\tilde{\Delta}f(x) = \tilde{\Delta}^2 f(x) = \tilde{\Delta}f'(x)\Delta x$$

La digression est:

$$\nabla f(x) = \frac{\Delta f(x)}{\Delta x} - \frac{df(x)}{dx}.$$

L'INTEGRATION

Commençons avec : $f''df=f'df'$, équivalent à: $df=\lambda df'$, où λ est le pas de la récurrence. Lorsque λ est simple, à savoir lorsque la condition : $d^n\lambda=0$, est accomplie, alors il existe un rapport analogique qui offre le lien entre f et f', et alors entre F la primitive de f et f.

Par exemple :

iii)   Avec: $d\lambda=0$; le calcul offre: $f''^2=f'f'''$; et par analogie: $f'^2=ff''$, $f^2=Ff'$

iv)   Avec: $d^2\lambda=0$; le calcul offre:
    $2f'f'''^2-f'f''f^{IV}-f''f''^2=0$; par analogie:
    $f(2f''-\lambda f''')=f'^2$, $F(2f'-\lambda f'')=f^2$

Lorsque λ n'est pas simple, le calcul conduit à se souvenir que: $f''\lambda'=f''-\lambda f'''$. Par deux voies, nous trouvons λ'' ainsi:

1) $f''^2\lambda''=-\lambda'f''f'''-\lambda f''f^{IV}+\lambda f'''^2$, qui conduit par analogie à :
$$\begin{cases} f^2\lambda'' + ff'(1+\lambda') = \lambda f'^2 \\ F^2\lambda'' + Ff(1+\lambda') = \lambda f^2 \end{cases}$$, contenant les cas: $\lambda'=0$, et: $\lambda''=0$

2) $f''\lambda''=f'''(1-2\lambda')-\lambda f^{IV}$, qui est futile pour l'analogie.

L'EQUATION ALGEBRIQUE A UNE INCONNUE

L'écriture générale de ces équations est bien :

$$\sum_{i=0}^{n} a_i x^{n-i} = 0$$

Où les a_i sont les données de l'équation. Pour la résolution, l'écriture suivante s'avère porteuse :

$$a_0^{n-1} \sum_{i=0}^{n} a_i x^{n-i} = (Ax + B)^n - C = 0$$

A B et C sont déterminés selon l'équation.

Pour exemples, prenons les cas de n=2 et n=3.

1) n=2 ; nous avons : $a_0 x^2 + a_1 x + a_2 = 0$; donc
$a_0^2 x^2 + a_0 a_1 x + a_0 a_2 = |Ax+B|^2 - C = 0$

Le calcul conduit à :

$$\begin{cases} A^2 = a_0^2 \\ 2AB = a_0 a_1 \\ B^2 - C = a_0 a_2 \end{cases}, \text{ donc:}$$

$$\begin{cases} A^2 = a_0^2 \\ 4A^2 B^2 = a_0^2 a_1^2 \\ C = B^2 - a_0 a_2 \end{cases} = \begin{cases} A^2 = a_0^2 \\ 4B^2 = a_1^2 \\ 4C = a_1^2 - 4a_0 a_2 \end{cases}$$

Et l'équation devient : $|2a_0 x + a_1|^2 = a_1^2 - 4a_0 a_2$

2) n=3 ; nous avons : $a_0x^3+a_1x^2+a_2x+a_3=0$, et :

$a_0^3x^3+a_1a_0^2x^2+a_2a_0^2x+a_0^2a_3=(Ax+B)^3-C=0$; d'où:

$$\begin{cases} A^3 = a_0^3 \\ 3A^2B = a_1a_0^2 \\ 3AB^2 = a_2a_0^2 \\ B^3 - C = a_3a_0^2 \end{cases} =$$

$$\begin{cases} A^3a_1^3 = a_0^3a_1^3 \\ 3a_1^2A^2B = a_0^2a_1^3 \\ 3a_1AB^2 = a_2a_0^2a_1 \\ B^3 - C = a_3a_0^2 \end{cases} = \begin{cases} A = a_0 \\ 3B = a_1 \\ (Aa_1 + B)^3 = a_0^3a_1^3 + a_0^2a_1^3 + a_0^2a \\ \quad + a_0^2a_3 + C \end{cases}$$

Globalement, la résolution de ces équations peut aussi être conduite sur la base des données suivantes :

$$\begin{cases} A^n = a_0^n \\ nB = a_1 \\ (Aa_1 + B)^n = a_0^na_1^n + a_0^{n-1}\sum_{i=0}^{n-1} a_1^{n-1-i}a_{1+i} + C \end{cases}$$

A chaque fois l'équation prend la forme :

$(na_0x+a_1)^n = n^nC$; donc:

$(na_0x+a_1)^n = a_1^n(na_0+1)^n - n^na_0^na_1^n - n^na_0^{n-1}\sum\limits_{i=1}^{n} a_1^{n-i}a_i$.

36

LA PROBABILITE PROGRESSIVE

Déjà, définissons ainsi le taux de certitude c:

c=

$$\frac{la\ fréquence\ de\ l'évènement\ e\ attendu\ dans\ l'échantillon}{le\ cardinal\ de\ l'univers\ des\ possibles}$$

$$=\frac{n_e}{card\{e,...\}}$$

Définissons ensuite le taux d'incertitude i tel que : i=1-c. maintenant, n étant le nombre d'occasions pour obtenir l'évènement attendu : le nombre de fois que la tentative est effectuée, la probabilité d'obtenir cet évènement est : $p=1-i^n$; elle croît avec n.

Puis, en souvenir que la probabilité de l'évènement contraire est : $\bar{p}$=1-p , nous trouvons :

1) la probabilité que les évènements distincts A et B surviennent ensemble est : $p_{A,B}=n_{A,B}\ p_A p_B$
2) la probabilité que A et B surviennent séparément est : $p_{A/B}=1-p_{A,B}$

3) la probabilité que A arrive avant B est : $p_{A;B}=$

$$\frac{p_A}{p_A + p_B} = 1\text{-}p_{B;A}$$

4) la probabilité de A selon la probabilité de B, est :

$$p_{A \supset B} = p_{(A;B),(A,B)} = p_{B;A}\,p_{A,B}\,.$$

LA VALEUR EXACTE DE π

Considérons la figure suivante:

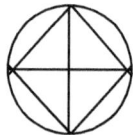

Il y a le cercle circonscrit et le losange inscrit ; la circonférence du cercle est parcourue selon la relation : s=Rα ; le losange en revanche est parcouru selon la relation : e=Rα$\sqrt{2}$.
Ensuite considérons cette autre figure :

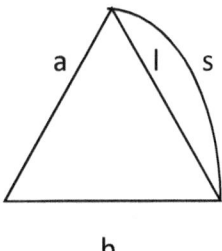

b

où :
$$\begin{cases} l = \sqrt{a^2 + b^2 - 2ab\cos\alpha} \\ e = \sqrt{|a - b|^2 + 8ab\alpha^2/\pi^2} \end{cases}$$

Le calcul montre que :

$$s = \frac{e + \sqrt{e^2 + 4R_m{}^2\alpha^2/\pi^2}}{2}$$; d'où nous tirons

précisément que : $\pi = \sqrt{2} + \sqrt{3}$; R_m est le

rayon moyen ; pour le cas du cercle, $R_m = R$, et :

a=b.

pour la mesure, l'unité de mesure de π est telle que :

$\pi = 3,14\ldots\breve{m}/\bar{m}$ (3,14…mètre courbe par mètre droit),

comme pour tous les angles euclidiens ou angles à

parois droits. Les conséquences sont :

1) le périmètre du cercle est la rotation du rayon

   selon l'angle de rotation ; donc :

   $$p = \int_0^{2\pi} R\,d\alpha = \pi D$$

2) la surface du cercle est le balayage du rayon

   selon le demi-périmètre ; donc :

   $$A = \frac{1}{2}\int_0^p R\,dp = \pi R^2$$

3) la surface de la sphère est le parcours d'une

   demi-circonférence sur une circonférence ;

   donc : $$A = \int_0^p \frac{1}{2}p\,dp = \frac{\pi^2 D^2}{4}$$

4) le volume de la sphère est le balayage d'une demi-surface de cercle selon une demi-circonférence ; donc : $V = \dfrac{\pi^2 D^3}{12}$.

LES REPRODUCTIONS STRUCTRELLES

Prenons simplement quelques cas :

$$C^{(a,b,c)}_2 = (a,b),(a,c),(b,c)$$

$$C^{a,b,c}_{x_i + x_j} = a + b, a + c, b + c$$
; la combinaison

des éléments a b et c selon la loi ou le régime :
$x_i + x_j$.

$$A^{a,b,c}_2 = (a;b),(a;c),(b;c),(b;a),(c;a),(c;b)$$

$$\Sigma_2^{a,b,c} = (a + b) + (a + c) + (b + c)$$

$$\prod_2^{a,b,c} = ab + ac + bc$$

Ces opérations sont extensibles pour toutes sortes de reproductions structurelles.

LES PUISSANCES OU LES EXPONENTIELLES

Observons que :
$$
\begin{cases}
2^n = \displaystyle\sum_{i=0}^{n} C_i^n \\
2^n = \displaystyle\sum_{\alpha=0}^{1} \left(\sum_{i=1}^{n} C_i^n \alpha^{n-i} \right)
\end{cases}
$$

Donc **a** étant un entier naturel, nous avons :

$$a^n = \sum_{\alpha=0}^{a-1}\left(\sum_{i=1}^{n} C_i^n \alpha^{n-i}\right) \quad = \sum_{\alpha=0}^{a-1}[(\alpha+1)^n - a^n]$$

En remplaçant n par x réel quelconque, nous nous souvenons au préalable que E étant la fonction entière, E(x)=E(e≤x<e+1)=e ; e est un entier relatif. Alors pour étendre les factorielles à tout nombre, les calculs offrent :

$$\begin{cases} x! = \displaystyle\prod_{i}^{E(x)-\lambda} (x - |\lambda|i) \\[2ex] \lambda = \dfrac{E(x)}{|E(x)|} \end{cases}$$

Alors précisément, nous avons:

$$a^x = \sum_{\alpha=0}^{a-1}\left(\sum_{1}^{E(x)} C_i^x \alpha^{E(x)-1}\right)$$

Plus encore, Chaque nombre naturel non nul est un complexe duomérique. En d'autres termes, chacun de ces nombres s'écrit :

$$N = \sum_{i=0}^{n}\alpha_i 2^{n-i} \quad = \prod_{j=1}^{m}\left(\sum_{i=0}^{q}\beta_i 2^{q-i}\right)_j$$

$(\alpha_i, \beta_i) \in \{0,1\} \times \{0,1\}$; n et q sont des latitudes duomériques, et m est le coefficient de complexité de N. N est premier si : m=1. Les nombres premiers sont donc les nombres continus ou saturés (solides), les autres étant les nombres discontinus ou insaturés (fluides). Autrement dit :

1) 2 est la souche (naturelle) de complexification
2) Tout nombre premier est racine de la fonction de complexité C définie par :

$$C(N)=\text{card } (N=\prod N_i) - 1$$

3) Tout nombre premier est alors un pieu de l'application de position P définie par : $P'(N)=C(N)$

Or : $\Delta\text{card } (N=\prod N_i) = \text{card}(N+x=\prod N_i) - \text{card } (N=\prod N_i)$.
Ici, $x= N-p$, p est le nombre premier nécessaire à N, et
alors: $\Delta\text{card } (N=\prod N_i)= \text{card } (2N-p=\prod N_i) - \text{card } (N=\prod N_i)$.

Par suite, $P''(N)= C'(N)=\dfrac{C(N)}{N-p}$. Autrement dit : les nombres premiers sont indépendants les uns des autres, à savoir qu'ils sont équivalents par rapport à tout repère auquel l'on les rapporte (toute remarque faite sur un nombre premier en tant qu'il est un nombre premier, est valable pour tous les nombres premiers).

www.ingramcontent.com/pod-product-compliance
Lightning Source LLC
Chambersburg PA
CBHW051822170526

45167CB00005B/2124